LINEA

Portrait of a Kaibab Squirrel

With Sketches of Other Wildlife on the
North Rim of Grand Canyon

By Joseph G. Hall

LINEA

Portrait of a Kaibab Squirrel

With Sketches of Other Wildlife on the
North Rim of Grand Canyon

By Joseph G. Hall

Designed by Amy Stewart Nuernberg

ISBN 0-9662734-0-0
Library of Congress Catalog Card Number pending

Manufactured in the USA
Pyramid Printing • Grand Junction, Colorado
Printed on recycled paper

On the front cover… a Kaibab squirrel surrounded by its habitat of ponderosa pine forest.

On the preceding pages… a topographical map of the southern tip of the Kaibab Plateau and the North Rim of the Grand Canyon.

Table of Contents

Discovery

It is fortunate, perhaps, that no matter how intently
one studies the hundred little dramas of the woods
and meadows, one can never learn all the
salient facts about any one of them.

— Aldo Leopold

For Betty,
Sally, Connie, Peggy and Lisa

Introduction

Linea was a real Kaibab squirrel. I first saw her on the Bright Angel Peninsula of Grand Canyon's North Rim one August and had my last sighting of her there a year later. During that interval, we were able to observe her much more intensively than any of the other squirrels encountered during thirteen seasons of study. What made this particular squirrel unique, and hence made her so easy to identify, was a short white line down the middle of her back. Her name, Lin´e-ah, is from the Latin word meaning "line."

In the field one morning watching for squirrels, an idea occurred to me. Wouldn't it be interesting to pull together the scattered facets of this squirrel's life and times into a coherent story? Surely that would provide a more engaging way for an interested visitor to learn about these squirrels and get an authentic taste of what summer on the North Rim is like than poring over a scientific treatise. These have a notorious reputation for stiff and formal prose, boring tables and obscure statistical notations. (Readers hardened to such fare are referred to my 1981 paper, *A field study of the Kaibab squirrel in Grand Canyon National Park*, Wildlife Monograph No. 75, pp. 1-54.)

This story developed from that morning's inspiration. It is like a stream whose flow is supplemented by a few tributaries. Episodes in Linea's life - events actually documented during the many hours of observing her - are the mainstream. But a major tributary consists of various nuggets of natural history revealed while watching other squirrels. These are worked into the narrative as if they had happened to Linea during a single season.

A smaller tributary is made up of vignettes of the natural histories from the lives of some of Linea's neighbors who shared the pine forest with her. Most of these have little if any direct impact on Linea but are part of the context of a Kaibab squirrel's life. Finally, trickling in here and there, are fictional rivulets that help add continuity and round out the whole. Although fictional in describing events not actually observed, they are factual in the sense that, based on reasonable inferences, what is described must have happened now and then during a typical squirrel's summer.

I have made a conscious effort to tell this story from the animal's point of view, presuming that it has some conscious thoughts and real emotions like our own. This is based on the many anatomical and physiological similarities we human beings share with other mammals, based on my own observations in the field, and based on the interpretations of other biologists.

It is a pleasant duty to acknowledge the support of the National Park Service, The Grand Canyon Natural History Association, The American Museum of Natural History, The Society of the Sigma Xi and The National Geographic Society. They helped make possible the field study which provided the biological information. The National Geographic Society allowed me to use part of their map, *The Heart of the Grand Canyon*, published in 1978.

My sincere thanks are extended to four colleagues who made valuable suggestions on improving earlier versions of the manuscript: Frederick R. Gehlbach of Baylor University, James P. Mackey of San Francisco State University, Chris Maser of Oregon State University and Jack S. States of Northern Arizona University.

I am especially grateful for my partnership with Amy Nuernberg of Pyramid Printing. Her editing suggestions, artistic eye and computer skills were indispensable in the content and graphic design. Rose Houk and Bruce and Frances Crowell reviewed the entire manuscript meticulously. Their editorial guidance resulted in substantial improvements.

My wife, Betty, devoted many a summer to camping out in a tent with small children and untold days cheerfully running tedious sample plots. She and close friends, Brenda Bechter Sabo and Kathleen McGinley, as well as Kathryn Daskal, Park Ranger, Grand Canyon National Park, gave me many suggestions for fine-tuning descriptions and narrative and I much appreciate their insight. The help and companionship of my brother, William B. Hall, during countless days afield were also major contributions.

The following donated time and talent in providing illustrations: Lisa Kearsley, our youngest daughter, drew the line drawings of Linea and the swift; Tom and Pat Leeson lent their superb portrait of a mutant white-bellied Kaibab squirrel; VIREO provided an image of a warbling vireo photographed by Brian E. Small; and NASA for their photograph of the moon.

Prologue

Splendid Isolation

The Kaibab squirrel, *Sciurus aberti kaibabensis*, is the most distinctive member of the group known as tassel-eared squirrels. It is of special interest as an example of evolution through geographical isolation. All the tassel-eared squirrels are strictly dependent upon a single tree species, the ponderosa pine. Several thousand years ago, these pines were much more uniformly distributed across the southwestern part of North America. Biologists infer that the tassel-eared squirrels were also more similar looking than they are today.

Then, as the climate gradually became warmer and drier, the pines at lower elevations disappeared, leaving the survivors as fragmented stands at higher elevations, scattered throughout the region. Of course the tassel-eared squirrels were isolated on these "sky islands" in like manner. Interbreeding between populations on neighboring highlands and plateaus became difficult or impossible because the lower intervening areas were no longer hospitable to them.

These isolated populations accumulated unique genetic traits over millennia of time, developing their own racial differences in appearance. Of these, the Kaibab form, found only on the isolated Kaibab Plateau on the north side of the Grand Canyon, is the handsomest with its white tail contrasting with a dark belly. The other races of tassel-eared squirrels, including the one as close as ten miles away on the South Rim of Grand Canyon, are collectively called Abert's squirrels.

It should be noted that this historical sequence is one of several proposed by students of evolution; there still remain numerous unresolved biological questions.

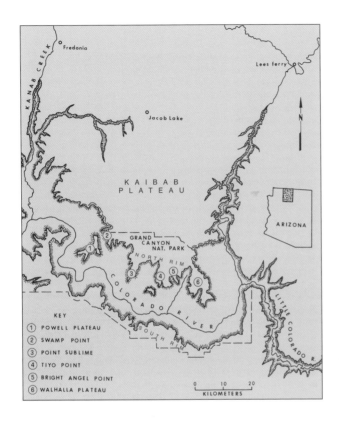

Above, map of the Kaibab Plateau in relation to
Grand Canyon National Park.

Right, what the eucalyptus forest is to the koala
of Australia, this open stand of ponderosa - or
yellow - pines on Powell Plateau is to the Kaibab
squirrel… provider of virtually all it needs.

Chapter One

Rhapsody in Green

No bird is more vocal or typical of aspen groves than the warbling vireo.

O n any morning in early June, hearing the song of the warbling vireo is as predictable as seeing the sun rise on Grand Canyon's North Rim. The hurried notes tumble out from high in the aspens and are part of late spring's biological signature of the Kaibab Plateau. This song is a small part, the largest part being the park-like empire of ponderosa pine that stretches out endlessly, acre upon acre.

© Brian E. Small / VIREO

Linea, a Kaibab squirrel, was also part of that signature. On this particular morning, she could hear a vireo singing but didn't listen; she was hard at work in the crown of a pine, gnawing off a branch tip from her swaying, precarious perch. What a beautiful but improbable white tail she had! And those fancy ears, for most of the year decorated by long tassels of fur that bend to the breezes. By now most of the tassels had been shed, a sure sign that summer was here. A maroon saddle lay astride the middle of her gray back. All in all, stunning adornment for a mere squirrel!

Of course Linea was oblivious to these genetic gifts. Yet the focus of her energy at this season was on ensuring they would be passed on to a new generation. The business at hand was construction of her nursery nest. Seizing a freshly cut twig with its cluster of needles, she nimbly galloped back to the trunk and worked it into the partly finished nest. About the size of a basketball, the nest was a mass of green needle-clusters jammed into the crotch where a large branch left the main trunk. The nest entrance was not at all obvious, but a closer look revealed the shadowed entry, squirrel-size, on the underside. Linea wasted no time admiring her work, scurrying out to another branch tip to repeat the routine.

A half hour later the nest looked finished. Held together by the bare-twig handles of needle clusters, such a compact structure is remarkably waterproof! However, Linea knew the inside needed a softer lining than the jagged twigs provided. Down the tree she went, and loped over to the aspen where the vireo was nervously flitting from twig to twig. She climbed to where the bark of a dead branch had lifted away from the wood. The bark had begun to decay into

In descending a pine, Linea displays the white line that suggested her nickname. Claws of her forefeet guide her direction while those of the hindfeet, now rotated backwards, control how fast she comes down.

bundles of shreddy fibers — ideal lining for a nest. Linea grabbed a mouthful of shreds and returned to her nest. Once inside, she pressed them into place in the floor of the chamber. Then back to the aspen for another load. A few more round-trips and she was satisfied — now the inside of the nest was safe and comfortable. The sun had risen higher, softening the chill of early morning. What better time for a bite of breakfast and a nap?

Breakfast for a Kaibab squirrel depends, like almost every need and activity in its entire life, on ponderosa pines. Day in and day out, season after season, the squirrel's world revolves about this one species of tree. So, when Linea got the breakfast urge, she headed for one of her favorite pines she knew would fill the bill. The nest tree, familiar as it was, wouldn't do. Now, you might suppose that one ponderosa is about as good as another. After all, they look almost as much alike as peas in a pod. Of course, there are the saplings, the middle-sizers and the grand old, yellow-barked patriarchs that give out an aroma of vanilla when basking in the sun. But beyond these age-related features, what can the eye distinguish? Probably not much besides shape and form. But Linea's nose took over

where her eye could not. She had learned what her ancestors several thousand years ago had learned for themselves: each individual pine has a unique "fingerprint" odor. All of them have the sugary inner bark she'd relied on most of the winter. But what varies are the ratios of bitter chemicals called terpenes. Why do these half-dozen or so substances vary so much from tree to tree? Linea didn't know. She did know how to tell, after a single bite through the bark, which trees were best tasting; she'd tested them all. On her mind's map she had plotted where these "feeding trees" were, and she was now on her way to one of them.

Her white plumed tail seemed to undulate, disembodied, across the brown forest duff because her nicely camouflaged body was relatively inconspicuous.

In a few minutes, Linea had reached the crown of an ordinary looking pine. Ordinary looking, that is, except for the crown itself which was obviously much thinner than most. Scrambling out to the tip of a branch, she neatly cut off a twig about three or four inches back from its cluster of green needles. Another slice and the cluster parachuted to the ground. This left the squirrel holding the bare twig, minus any encumbering needles. Back on firmer footing near the trunk, Linea squatted, holding this three-inch "pencil" in her paws. She turned it deftly, corn-on-the-cob fashion, rapidly flaking off the brown outer bark and eating the spongy, red inner bark beneath. It was saturated with sugars manufactured by the needles just discarded, with only a hint of objectionable terpenes. But in this particular pine they seemed less noticeable than most.

She was soon down to the bare, yellow, wooden core and let it drop to the ground. By chance, it came to rest alongside the same needle cluster which, through photosynthesis, had produced each of its woody cells. For several yards around, the forest floor was littered with similar clusters and yellow twigs, infallible clues to that tree's invisible chemical attraction to Linea and her kin.

Thirty minutes and a dozen twigs later, Linea's appetite was dulled. She gazed vacantly through the crown, a little thinner than those of neighboring trees due to her repeated harvesting of twigs. High above, two ravens soared against the backdrop of a few puffy cumulus clouds. A tree squirrel, spending all its days here, still could not know what a spectacular, sweeping overview of Grand Canyon's temples and buttes, cliffs and terraces, side-canyons within side-canyons the ravens saw below. Their eyes saw all the details with pinpoint clarity but, irony of ironies, raven brains are not programmed to appreciate the ensemble as scenery!

As the day warmed, a mood of languor crept through the forest. The vireo's compulsive singing had waned, then stopped altogether. Linea stretched out on her black belly, tail hanging to one side of the branch. Her attention shifted to a troop of pygmy nuthatches as they flitted from branch to branch of her tree. Working up, down and sideways, they piped trivial nuthatch gossip to each other with high enthusiasm. Then they were off; Linea's eyelids drooped, opened briefly, drooped again and closed.

When feeding on the inner bark of twigs, high in the crown of a pine, Kaibab squirrels discard two telltale clues that drop to the ground - fresh, green needle clusters and cleanly stripped yellow twigs.

Chapter Two

Dramas at Harvey Pond

About one week before birth, baby squirrels show tiny claws on each foot. The normal gestation period is about six weeks.

For the past couple of weeks Linea had felt an increasing heaviness. Climbing, and especially jumping from branch to branch, seemed to require more and more effort. So she had become less active, spending more time in the nest. Then, suddenly late one afternoon, painful cramps began and grew stronger. With irresistible waves of pushing, she gave birth to four squirming, pink babies. Immediately she licked each one clean, then consumed the placenta. Dark organs were barely visible through their transparent skin. Each foot was tipped with tiny black claws, hints of future climbing ability. The most striking features were the two large, blue bulges on each side of the head - eyes - which would not open for a full month. Perhaps least striking were the undeveloped ears, mere depressions at this stage. And at the other end was a ridiculous little naked pink whip of a tail. It seems inconceivable that each of these vulnerable, wormlike packages of warm protoplasm contained the microscopic blueprints needed to guide it in its development into a bright-eyed, fully-furred forest sprite that can know enough and learn enough to maintain itself, escape enemies, find a mate, and produce more of its kind. Yet we know that it does. Incredibly, in developing from fertilized egg to their present form during the previous forty days, the embryos underwent transformations incomparably more radical than those lying ahead. Similar feats are accomplished by all the other members of Linea's neighborhood. How many tiers of miracles can a forest of yellow pine, a relatively simple, often sleepy looking ecosystem, contain?

Never to be burdened by that question, the four wriggling newborn soon found nipples and had their first nursing.

As June came to an end, the young squirrels had prospered. Eyes and ears about to open, bodies covered with a thin coat of fur, and their birthweights tripled, their demands on Linea increased. To provide enough rich milk, her own food intake had to rise to the occasion. The simplest option was simply to step up her consumption of inner bark, an omnipresent but not especially rich food item. A more efficient option depended on the availability of female cones. A complex of weather conditions acting over a two-year span dictates whether the ponderosas on the Kaibab Plateau will have a good, fair, or poor crop of green cones in a given spring. Fortunately, this year the pines were burdened with a bumper crop. Hidden under each cone scale were two plump seeds, much more nutritious than inner bark. So, shortly after the debut of the quadruplets, Linea phased out of her bark diet and into one of pine seeds.

As violet-green swallows whistled in another dawn, Linea set out on a breakfast excursion. Leaving the four youngsters asleep in the nest, she made her nimble way up the trunk, then out to a branch tip. From here a modest leap carried her to the crown of a neighboring pine. Then, down its trunk and to the ground. A casual observer would probably have concluded, with some smugness, that descending the nest tree directly, rather than by her indirect route, would have saved her both time and energy. But when such simple solutions to life's problems are not followed, there usually is some reason. Could some long-term benefit result at the cost of short-term inefficiency? Further observation reveals that nest trees are rarely isolated. Typically they are surrounded by several neighbors, close enough to make aerial transfers easy for a squirrel. Is it possible that a potential predator might be "tipped off" about the location of a nursery nest more easily if the mother always climbed up and down the nest tree? No one knows for sure, but the idea holds appeal.

A stocky little horned lizard skittered out of the way as Linea cantered across the needle carpet of the forest. Off to the left, something white caught her eye. An antler of a young mule deer buck, shed last March, lay bleaching in the sun. A lucky find for Linea, for this concentrated store of calcium would enrich her milk. So she devoted the next ten minutes to gnawing the antler tines. She was suddenly frightened by the fast-moving shadow of a large bird, gliding just below tree-top level. After a mad dash behind the base of the nearest tree, a glance upward assured her it was only a raven. Such a false alarm, bearing an energy price tag, was another example of short-term inefficiency. But prudence had become habitual when it came to large moving

Constructing a rack of solid bone antlers each year is a major nutritional job. But it benefits the overall vigor of the deer herd by favoring mating by stronger, more competitive males. It also contributes to the supply of calcium available for squirrels, porcupines and other rodents in the community.

shadows. They could be cast as easily by ravens as by goshawks which are major predators of Kaibab squirrels. Finally, with the raven out of sight and her heartbeat slowing to normal, Linea flicked her tail, surveyed her surroundings, then climbed high up in the pine. Green cones, in bunches of three or four, hung from almost every branch tip. Eagerly, she climbed out to a pre-carious-looking perch near the end of one branch. With a neat bite she cut a cone from its base and carried it back to a steadier seat next to the trunk. Now her procedure was reminiscent of stripping bark from twigs — she held the cone in both paws and skillful-ly chewed away each scale to expose the two juicy seeds beneath. One by one, the scales rained to the ground, finally followed by the yellow core. No more work per minute than harvesting inner bark, but the reward was richer. Over and over she repeated this routine, ignoring a brash western tanager that lit nearby to contest her presence. A half hour ticked by and the sun began to filter its light over the shim-mering Painted Desert far across Marble Gorge. Linea's appetite was satisfied and for a few seconds she crouched motionless, soaking up the solitude. Then she realized she was thirsty.

The sponge-like Kaibab limestone had soaked up May's snowmelt so completely that almost no standing water could be found. The topography and geology of Grand Canyon dictate that the bulk of precipitation falls as rain and snow at high elevations on the Kaibab Plateau's surface. But the water is only reluctantly available to resident animals. Water percolates through several thousand feet of porous rock until stopped by a nonporous layer. At Roaring Springs or Thunder River, far below the rim, it emerges in great cascades, creating oases in the Inner Canyon desert. But these are inaccessible to the animals of the cool conif-erous forests high above.

Luckily, there are widely scattered sinkholes in the Kaibab limestone. Their floors long ago became clogged with silt that prevented percolation. Thus, they persist as small ponds. Few and far between, the location of each of these "pearls of great price" is common knowledge to creatures within a large radius. Linea now headed toward the nearest one, Harvey Pond. On the way she passed by a patch of blue lupines, the most common flowers in the forest

The "wing" on each pine seed is its mechanism for being spread far and wide in the wind. But the squirrel's feeding strategy interferes - it harvests the seeds before the cone is mature enough to release them.

The drops of water along the leaflets and in the center of each lupine leaf are a boon to small animals in the dry season. Notice how each drop acts as a miniature magnifying glass, revealing tiny surface hairs.

openings. Beyond the beauty of their showy flowers, lupines have a unique role in providing water to small animals. Each spoke-like leaflet is creased along its length and converges to the center of the leaf. Consequently, water from recent rains or overnight condensation of dew accumulates as a single large drop. In addition, the leaflets are covered with tiny hairs that may help to retain water. The lupines act as miniature, ephemeral reservoirs. Linea ran to the patch of blue and carefully lapped up jewels of moisture from each leaf.

Her thirst, now at least partly satisfied, she resumed her journey to Harvey

Pond. As she approached the quiet little sink, tucked up against a limestone outcrop and a steep forested hillside, she felt a strange uneasiness. Instead of making her usual direct approach to the water's edge, she elected to climb a twenty-foot sapling, the better to reconnoiter. No sooner had she reached a convenient perch than a slight movement in a heavy clump of firs caught her attention. Turning her head slowly, she made out the form of a bobcat emerging from the low branches into full view.

It glided, graceful and cautious, down to the pond, stopping every few feet to survey the scene. Directly in its path, Linea's pulse began to race and she froze on her

Prince of predators! This bobcat pausing to drink at Harvey Pond is a living pinnacle of adaptation. Its every feature - anatomical, physiological and psychological - is honed to its keenest edge for the challenging life of a carnivore.

perch. A study in stealth, the bobcat paused at the foot of the tree in which she crouched. Its great padded feet had made no hint of a swish or rustle on the dry twigs and needles. It turned its head left, then right, and Linea could see the big yellow eyes, vertical pupils now reduced to slits, scanning the shore of the pond. Fortunately, it did not bother to look up. Beautiful, black-tufted ears were cocked alertly to catch any telltale sound made by a careless meadow mouse. Linea dug her claws even deeper into the bark of her branch and moved not a hair. Then, with a reflex twitch of its "bobbed" tail, the bobcat eased down the final slope to the water's edge, stood erect momentarily, then crouched to drink. A Williamson's sapsucker called raucously from the wooded hillside, then brazenly swooped down to the water's edge opposite the bobcat. Seeing its mistake, it hastily flew off again without drinking. The bobcat did not interrupt the steady rhythm of its lapping. Like all successful predators, it had long since precisely reckoned the likelihood of potential meals — the sapsucker clearly qualified as unlikely. In fluid motion the bobcat rose to its full height, glanced quickly around the shore, then turned to depart. Linea's last impression was the bold black and white markings on the backs of its ears.

Relaxing at last, Linea gave a flick of her tail, then slapped her forepaws against the branch and barked softly: "Chuck, chuck!" But, for the moment, she elected to stay put. Such precautions cost nothing and often prove invaluable. Immediately the sapsucker returned, followed shortly by a pair of juncos. All drank and bathed in the shallows as if the bobcat had never existed. That was good enough for Linea. She scrambled down and was off to the pond for her long-awaited chance to drink. At the shore line, Linea returned the gaze of a bright blue damselfly on a blade of grass a few inches away. The gossamer-winged insect was now undeniably as fit for a terrestrial way of life as was the tree squirrel. Yet, less than an hour ago, in its nymphal stage, the damselfly had been an active member of the aquatic world just beneath Linea's chin. Though Linea could not see it, feverish activity was going on in the pond. At this time of year, the major actors in the drama were larval tiger salamanders. Fish-like in general form, they nonetheless had other unique features. Four small but distinct legs and a pair of large, feathery gills attached to the neck identified them as amphibians. Scores of them swam leisurely among floating algae, powered by undulations of their long, translucent tails.

Whenever one relaxed its tail, it sank slowly to the muddy bottom, spreading its legs as stabilizers. Small worms, snails, mussels, insect larvae, and crustaceans were teeming in the water, and the salamander larvae were busy feeding on them. Several of the salamander larvae were, in turn, being fed upon by predatory diving beetles that attached themselves like miniature leeches. One larva noted a stubby, pink object alternately breaking through the surface, then disappearing. It swam up to the surface to investigate the potential food. But it was merely Linea's tongue.

The most bizarre thing about these larvae was that some had much larger heads than the others. However, none of the salamanders showed intermediate head sizes. The large-headed ones were not a different species or in a later stage of growth from the small-headed ones, but different larval types destined to develop into identical adults. The big difference in appearance is reflected in behavior — the large-headed forms cannibalize the small-headed ones, except for their close relatives. How they manage to discriminate degree of kinship, however, is a fascinating unknown.

Although Linea satisfied her thirst here, she had no inkling of the fastpaced drama of the pond. Its water was a crucial element in her survival, but its community of life was not part of the world she knew.

Tiger salamanders hatch two types of larvae: broad-headed cannibals and narrow-headed herbivores. What advantage this might give is not yet understood.

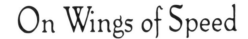

Chapter Three

On Wings of Speed

Linea's four youngsters were becoming stronger and more adventurous these days. It seemed like only yesterday that clambering unsteadily about on the branches near the nest was the most daring of exploits. Now, weaned and as well coordinated as their mother, they enjoyed acrobatic games of follow-the-leader all over their home tree. Although indisputably Kaibab squirrels, they were not just smaller, carbon copies of Linea. For one thing, they had distinct ear tufts, a feature adults lack during the summer. In addition, their tails were tinged with gray and sparsely furred in contrast to their mother's white, plumed flag. Finally, their boisterous sociability seemed out of character compared to her staid, even antisocial, demeanor.

One of the males, named Mumbles for his habit of grunting and chuckling softly whenever Linea was near, had a favorite play routine. It involved scurrying wildly about on the ground, then suddenly running full tilt at the nearest tree trunk and ricocheting off it like a rubber ball. Occasionally he tried to entice Linea into joining the fun, but she stolidly ignored him.

Near the top of a tall fir in a ravine nearby, a Cooper's hawk had its nest. Three eggs had hatched there a few days before Mumbles and his sisters were born. Gradually, the fledgling hawks and squirrels became aware of each other. One morning, one of Mumbles' sisters had just descended the nest tree and was walking deliberately across the duff. Without warning, one of the newly fledged Cooper's hawks dived on her at a steep angle. Giving a frightened squeak, the young squirrel nimbly doubled back to the nest tree and dodged to the opposite side of the trunk, with time to spare. The thwarted hawk veered off to one side, none too gracefully, but just in time to avoid colliding with the tree. It seemed that almost any moving target triggered the instinctive impulse for pursuit in the young raptors; the scenario was repeated almost daily.

Linea and her family were well endowed with an innate fear of large birds in flight. But their escape maneuvers developed a certain air of becoming routine. Then one hot, drowsy midday, as grasshoppers buzzed lazily and the squirrels were taking a welcome siesta, disaster struck. Banking and weaving swiftly through the forest, no more than six feet above ground, an adult goshawk burst on the scene. It was reminiscent of the Cooper's hawks though distinctly larger and mantled with a dark gray rather than brown. Fierce red eyes were set in a face marked with bold black and white stripes, giving it an implacable expression. Sensing this to be no ordinary threat, the squirrels were galvanized into action. In a pandemonium of claws raking off flakes of bark, most of them careened around to the lee side of the nearest tree. But the goshawk had unerringly targeted Mumbles, splayed out on his stomach on a sunny limb. He had barely

When on the ground, Kaibab squirrels must be quick to dodge around to the protected side of a pine trunk in times of danger.

begun his scramble to safety before the goshawk swept him away in her talons, hardly pausing in flight. She disappeared with her furry booty which gave one final scream. Silence. The remaining squirrels clung to their perches, frozen in place . . . even the leaves of the lone aspen hung motionless. Several minutes elapsed. A grasshopper broke the oppressive stillness and resumed a tentative buzz. Then another and still another added to the chorus.

Little by little the squirrels resumed their normal behavior - greatly subdued to begin with and much more cautious. By a half hour's time they seemed completely recovered. Whether they all actually knew what had happened or had learned by the experience would be hard to say. High, high above, utterly oblivious to the drama just enacted below the canopy of ponderosas, were other aerial predators, the white-throated swifts, uttering rapid, high-pitched twitters on the wing. They pursued their own prey — flying insects of many varieties. Most daring of avian acrobats, they are sometimes confused with swallows. But when compared with them, graceful in their own right, swifts are simply in a class by themselves. When, as often happens, they are seen flying together along the cliff faces of the Canyon's rim, the swallows seem the embodiment of prudence, always under control. Swifts, on the other hand, fly with what looks like complete and reckless abandon, throwing caution to the wind.

Imagine the terrifying thrill of climbing into the cockpit of a swift's brain . . . seeing the world through its eyes on one of its death-defying sorties! What a giddying vertical plunge it would be to catapult from the brink of the Canyon at Bright Angel Point, down past the face of the Kaibab limestone, past the even more sheer-walled sculpture of the Coconino sandstone at speeds well above 100 miles an hour. (At this velocity its wings cut the air with force enough to make a sound like that of "ripping a bolt of silk," as one observer aptly put it). Then, leveling off again near the top of the Supai formation, zigzagging an erratic course to snap up hapless insects, and regaining the lost altitude by power strokes of the narrow, swept-back wings (wings that give the appearance of beating alternately rather than in unison). What sublime panoramas would unfold as this diminutive black-and-white comet sped west past Oza Butte, over the saddle between Widforss Point and Buddha Temple, noting the shifting forms of Cheops, Isis, Horus and Osiris Temples rising between it and the Colorado River. Then, swiftly over the Colonade with

Tiyo Point to the right and flat-topped Shiva Temple to the left. (Shiva, with its two isolated stands of ponderosas, was probably once inhabited by its own tiny population of Kaibab squirrels, snuffed out by some quirk of climate or genetics 10,000 years ago.) Finally, after even more kaleidoscopic visions, what a climactic finale it would be in late afternoon to be flung headlong toward a narrow, vertical crack in a cliff, to decelerate to zero miles an hour with exquisite precision that placed the bird safely inside its dark, protected roost. Could there possibly be an equal to that extravagant an adventure of the senses? If the song of the canyon wren, with its clear cascading notes, personifies Grand Canyon for the ear, the sight of the white-throated swift, demonstrating ultimate mastery of flight, personifies it equally well for the eye, in this expansive theater, presumably created especially for it.

The white-throated swift is Grand Canyon's most daring aerialist. As it pursues insects on the wing it gets perspectives such as this one of Shiva Temple. Isolated from the North Rim, Shiva is similarly crowned with acres of ponderosa pine.

Chapter Four

Competition for Seeds

Some bird families can be identified by posture. Flycatchers, like this western wood-pewee, for example, always sit on a perch rather than stand on it as a robin would. They typically perch out in the open, affording them an unobstructed view of passing insects.

In the gray half-light before dawn, a western wood-pewee cheerfully proclaimed the advent of a new day with his mirror song, "pee-whit...whit-pee." Perched on a broken off pine stump a few feet above ground, the bird flipped its tail up and down slowly as it called and turned its head about, watching for insects. Towering next to the stump stood an aspen and on one of its horizontal limbs huddled the pewee's mate, brooding young. The nest was so neat and compact it appeared to be a fibrous outgrowth of the limb itself. Halfway down the trunk was the entrance to an abandoned wood-pecker's nest. It served now as the nursery of Chickaree, a red squirrel, and her three offspring. Awakened by the pewee's repetitious matins, Chickaree stirred uneasily, then climbed up high enough to poke her head out the round doorway. Too early! Back down she clambered to the warm, furry mass below... the new day could wait for a few more minutes. Rays of sunlight broke over the lip of the Painted Desert; they streamed across the darkened abyss of Marble Gorge and filtered through the yellow boles of the pines crowning Walhalla Plateau. Soon Chickaree could detect the vanilla scent given off by sun-warmed ponderosa bark and could hear the faint "hip, three cheers!" of an olive-sided flycatcher singing from atop a Douglas-fir just over the rim. She scampered off for breakfast.

Ordinarily Chickaree was content to live among the spruces and firs of higher, cooler elevations. But this was not an ordinary summer. Somehow she'd gotten wind of the bumper crop of pine cones below and moved down to take advantage of it. This morning she dashed down the aspen and up to the crown of the nearest pine to harvest the plump, green cones. It so happened that Linea was busy at the same task in a neighboring pine, a situation that upset Chickaree. Her territorial instinct was stronger than Linea's and, despite being a seasonal interloper, she resented any competition. So she went about her business with frequent sidelong glances, nervous flicks of her tail, and a few chattering scolds thrown in for good measure. Her method of harvest was quite distinct from Linea's. Instead of consuming seeds one by one, on

the spot, Chickaree cut off the terminal needle cluster, then each cone from the branch tip. The cones came plummeting to the ground where they landed with a thump. After a couple of dozen had fallen, Chickaree raced down to retrieve them, carrying each in her jaws to a nearby midden at the base of a ponderosa. Now and then she could not resist peeling the scales from a cone or two and eating the seeds. The bulk of her harvest, however, she carried intact to her midden, a routine not lost on Linea, among others.

A white-breasted nuthatch worked her way, headfirst, down the trunk of the midden pine. She looked from side to side and, with occasional nasal chirps, probed here and there for insects hiding in bark crevices. When she reached the midden, the nuthatch poked about in the pile of cones, hoping to find one with scales soft or loose enough to yield to her sharp bill. No such luck — a few more nasal chips of disappointment. But she did notice Chickaree enjoying a snack on the forest floor not too far away. She promptly flew closer, just in case there might be a chance to glean some leftovers; the nuthatch was well aware of Chickaree's untidy feeding habits. Shortly, a passing Uinta chipmunk, alerted by the

Like the Kaibab squirrel, the Uinta chipmunk in Arizona occurs only on the Kaibab Plateau. They are excellent climbers. Golden-mantled ground squirrels, often confused with chipmunks, rarely climb, are much plumper and lack head stripes.

activity of Chickaree and the nuthatch so close together, decided a little reconnoitering of his own could do no harm.

Thus the stage was set for a four-link chain-reaction triggered by Linea. Having had her fill of pine seeds, Linea began a deliberate descent to the ground. Chickaree, still nursing resentment over her being in the neighborhood in the first place, dropped a few tidbits to focus full attention on her larger cousin. No sooner had Linea reached the ground than Chickaree launched a noisy attack, sure that her midden was now under direct threat of invasion. Each bound forward was punctuated by a burst of

staccato scolds. Not in the mood for confrontation after her meal, Linea made a perfunctory dodge or two, then cantered slowly along with the agitated red squirrel pursuing as closely as prudence allowed. Meanwhile, with Chickaree temporarily offstage, the nuthatch was quick to flit down to the ground. She greedily began to snatch up leftover seed fragments. But the chipmunk was only a few seconds behind her. Confident that she would yield, the chipmunk made a rush directly toward her, only to stop dead in his tracks! What was this? Without warning the prim little bird had undergone a startling metamorphosis. She fluffed out her body feathers, half-spread her wings and lowered her head and neck as she wheeled around to face him. Now appearing twice as big as before, the nuthatch swaggered from side to side, bobbing, feinting, and weaving like a boxer, tipping now to the left and now to the right. Transformed into a formidable-looking antagonist, she parried each of the chipmunk's tentative charges. The astonished chipmunk was at a complete loss and, with a squeak of disappointment, suddenly turned and darted for the cover of a fallen log. Regaining her former shape, size, and demeanor, the nuthatch picked up the last few crumbs, then flew off to the aspen grove.

Chapter Five

The Predatory Way

Pewee shading nestlings

Two hours after the nuthatch's triumph over the chipmunk, the sun had risen well toward its zenith, and late morning heat was slowing down most activity in the forest. A grasshopper, flushed by a meandering doe, flew noisily to another grassy pocket. The female pewee had given up brooding her nestlings and now stood over them with bill open and wings partly spread to provide shade. She chanced to catch some movement in the leaf litter beneath: a sinuous black, brown, and cream chain of a pattern. Undulating slowly along the ground was a gopher snake. The pewee stiffened in alarm, drew in her wings slightly, and froze in position, following the snake's advance below her. Every few seconds the snake extended its

forked, black tongue, flicked it up and down, then drew it back into its mouth. When the tongue was extended, microscopic scent particles were trapped on its wet surface. After withdrawing it, the snake pressed the tongue's forked tips into two sensitive pits in the roof of its mouth. Nerve endings in the pits, stimulated by scent particles, relayed odor messages to the snake's brain. Suddenly, the snake stopped in response to a "taste" of young red squirrel scent. It looked about with care but got no clue on location since it was on the opposite side of the aspen where Chickaree's young huddled in their nest.

As luck would have it, Chickaree came running toward her nest, having just chased Linea away from her midden. She didn't notice the motionless snake and climbed nimbly up the aspen to her home, pausing momentarily at the entrance to survey the area. But the snake crawled quickly around to follow Chickaree, arriving in time to see her tail disappear into the nest hole. This

This gopher snake, like all snakes, is completely carnivorous and stone deaf. But it has good eyes and a very keen sense of smell which help it locate small birds and mammals, such as this frightened chickaree.

was all the snake needed to locate the source of squirrel scent. He immediately began to climb, an apparently impossible task due to the aspen's smooth trunk. Nevertheless, there were just enough scars and knots to allow the infinitely flexible body of the snake to get a hold here and there. It was the back edges of the belly scales that actually gripped these small irregularities in the otherwise seamless bark. The slight scraping sound of scales on bark caught Chickaree's ears, however, and she scrambled up to look out of the door. What she saw threw her into an instant panic. With a stream of chattering, she dashed down the trunk to confront a predator easily capable of consuming an animal her size. The first pass she made was totally ignored, and the snake continued its slow but steady upwards contortions. In desperation Chickaree made a second, even bolder pass, which was met by a loud hiss. On the third and final pass, she took the ultimate risk and struck the snake hard enough to make it lose its grip and tumble to the ground. The snake, apparently re-evaluating the situation, gave up and slithered off to some logs bordering the glade.

Both Chickaree and the pewee watched from the aspen. Squirrels and songbirds almost certainly cannot express relief on a conscious level. Perhaps they do not even feel this emotion. For a reptile to be able to express frustration seems even less likely. So the gopher snake, unblinking, impassive opportunist, lost no time searching farther afield for a meal. Eyes alert and tongue flicking, he soon picked up the scent of a robin foraging among low grasses. The snake began the approach, indirectly and with great care. Part of his strategy was to alternate short, slow advances with long periods of frozen inaction. During one interval the robin made the fatal mistake of wandering inside the snake's striking radius. Fast as a robin's reflexes are, they were not fast enough to avoid the snake's blinding strike. In an instant the snake had clamped its jaws onto the robin's neck and threw two muscular coils around its body. Suffocation did not take long. The snake released its grip on the robin's head, which fell limply. With its two coils still wrapped around the bird's body, the snake took the robin's head in its jaws and began to swallow whole an animal whose greatest diameter was at least four times larger in diameter than itself. The head posed no problem. But once that disappeared, several unique anatomical features were called into play to allow the snake to engulf the rest of the robin's body. One

Without extremely flexible skin it would be impossible for a snake to swallow this robin which is much wider than the snake's own mouth! Its extra strong digestive juices make up for the fact that the teeth of snakes are designed for moving their food but not chewing it.

anatomical advantage was the ability of the right and left lower jaws to unhinge from the upper jaw and to work independently of each other. Thus the snake could "walk" the prey into its mouth. Another indispensable feature was a long flexible extension to the windpipe. When extended out of the mouth like a snorkel, the snake could breathe freely when it would otherwise be impossible. Finally, incredibly elastic skin was crucial to success. As the bulkiest part of the robin was pulled inside, the skin of the snake's head and neck was stretched so much that the scales, normally resembling overlapping roof shingles, looked like separated flat warts attached to a thin membrane. The snake now presented a grotesque, misshapen image of its former self.

After the robin's body moved through the throat, the snake's head and neck returned to normal. Out came the tongue for the first time in minutes and the snake miraculously regained its calm, expressionless appearance. Its large, swollen midsection was mute testimony to an amazing feat. Laboriously, the snake retired to the shade of a bracken fern to allow digestive processes to take over.

Chapter Six

Ordeal by Live-trap

L inea had been exposed to people, from time to time, all her life. Occasionally hikers would pass through her groves, usually not seeing her at all since she spent most of her time in trees above their line of sight. Her standard procedure, soon imitated by the three youngsters, was to freeze in position, almost guaranteeing that she would go unnoticed. It was a different situation, however, when she was surprised on the ground. Then her showy white tail was conspicuous against the brown pine needles, and people often would shout and point, then make a direct approach, camera at the ready. She had only to race for the nearest pine, climb to the lower branches on the side opposite her pursuers and hope they would go on their way. In case they were more persistent, her tactic was to climb up to the crown and hide among the needle clusters. Here her white tail was surprisingly difficult to see when sunlight reflected off the shiny needles surrounding her. The noisy humans would shortly abandon their efforts and head off along the trail. They were never a source of serious concern — that she reserved for a coyote, a bobcat, and especially an airborne threat such as a goshawk.

But for the last couple of weeks she had become aware of one person that she encountered more frequently than stray hikers. Slightly-built and armed only with binoculars and notebook, he might be seen slowly making his rounds at almost any time of day, even early enough to record her emerging from her nest. Unlike the typical hikers, he apparently wandered more or less aimlessly about her territory and would sometimes spend periods of time simply observing her. Although she always kept her distance, she eventually came to perceive him as non-threatening.

One morning she noticed him placing several small objects among the trees. Her curiosity was tempered by suspicion so she waited until after he'd left to investigate. At the nearest site, which she approached with some care, were various food tidbits. The pinyon nuts were familiar, but the walnuts, peanut butter-covered crackers, and orange slices were new to her. Human scent was strong on everything and persuaded her not to sample. The other bait stations she approached were like the first and she sampled nothing. The last station she visited was not far from Chickaree's main cone midden and she followed an impulse to detour to it. As usual, Chickaree was keeping a sharp eye on the midden. As Linea headed in its direction, Chickaree came to its defense chattering a torrent of abuse. This time, however, Linea showed no sign of yielding, and her subtle body language persuaded Chickaree to rein in her charge short of contact. For her part, Linea ignored the theatrics, picked out one of the green cones from the midden and loped off through the forest with it. Late that afternoon as Linea was returning to her nest, she surprised a golden-mantled ground squirrel helping itself to the nuts and crackers at one of the bait stations. Although she was tempted to do the same, discretion again intervened.

Over the next week the biologist maintained his morning routine of replenishing the bait, noting which stations and which baits were most popular. By mid-week Linea had watched both of her own daughters sample the bait and her resistance weakened.

(Linea's third and most independent daughter was seen with her family less and less. Finally she disappeared altogether, either having fallen prey to some predator or having wandered off to another area on the plateau.) Linea gingerly sampled the offerings at a few sites. Seeing the rock squirrels boldly taking bait at the two stations closest to the edge of the Transept side canyon, gave her even more confidence. Orange slices, although a completely unnatural dietary item, were Linea's first choice. Perhaps during this dry season their moisture and high sugar content made them prime delicacies. In any event, with each passing day Linea showed less hesitation. She even became bold enough to head toward a newly baited station before the biologist was out of sight.

Then one morning, perched impatiently close to her favorite bait site, Linea watched the biologist place a strange, wire mesh box near the fresh bait. Having had no experience with live traps, she was only mildly suspicious. Soon she was busy as usual, feasting on the orange slices and walnuts. The next morning she found the bait just outside the entrance to the trap and the following day the biologist locked the door of the trap open and placed the bait on

The same size as Kaibab squirrels, rock squirrels prefer open, rocky areas rather than forests and seldom climb trees. For these reasons, they are not important competitors for food.

the floor just inside. Step by step and day by day she became more accustomed to the trap, finally going completely inside to retrieve bait from behind the treadle without a second thought.

What would be Linea's most traumatic day of the season dawned bright and clear. Confidently, she approached the baited trap site as soon as the biologist had disappeared

The glass jar is necessary as it allows biologists to monitor the squirrel's reactions very closely and thus avoid giving too large a dose of anesthetic.

The squirrel, now fitted with a collar, is coming out of anesthesia and is almost ready to be released.

into the forest. She entered the trap to reach for the bait in the back and, as usual, stepped on the treadle. She had no way of knowing that the door was not locked open this time and hardly noticed when it fell shut behind her. Turning to leave, she found her exit blocked. Suddenly it dawned on her that she was trapped! Her accustomed world still surrounded her but she had lost access to it. Frantically she threw herself against the walls of the trap and rattled the doors at each end. Finally she settled down at one end, exhausted, and anxiously waited. The trap had intentionally been placed on the northwest side of a large pine where it would remain shaded for at least two hours, so Linea was in no immediate danger of heat stress. Not a breeze stirred through the sunny, silent stand of ponderosas.

About a half hour after the initial crisis of being trapped, she caught sight of the biologist and his assistants coming directly toward her trap, speaking softly. In a resurging panic, she clawed and bit savagely at the unyielding steel mesh. It was to no avail. With her pulse racing, she looked about in desperation for some possible escape. What followed were stages of a chaotic nightmare. Attempting to minimize her fright, the biologists kept as many tree trunks as possible between Linea and themselves as they approached. At the last moment, one of them stepped out from behind the closest tree and quickly dropped a dark green shirt over the trap. Linea barked in alarm but, no longer able to see beyond her prison, soon calmed down. Next she was shooed from the trap into a holding bag. Finally, the critical transfer was made from the bag into a large jar in which was a cotton ball doused in anesthetic. At first she rushed madly around inside the jar but within less than a minute became dizzy. The faces of the biologists and the forest beyond them began to swim in circles and, losing her balance, she fell. She groggily climbed back to her feet, only to fall again. This time she lay on her side, twitching a bit, unable to rise. Exactly the sign the biologists had waited for! They immediately removed her from the jar. While one of them held the warm, limp body, the other adjusted a radio-collar around her neck — loose enough to avoid choking, but tight enough to stay on. Then the collar was clamped closed, and Linea was placed gently on the ground at the base of the big pine. The biologists backed off a few yards to watch. Gradually the effects of the anesthetic wore off and Linea's head began to clear. The forest around her "settled down" and she saw the two biologists sitting quietly at a distance. After a minute or two she felt completely recovered. Following an instinctive impulse, she jumped to her feet and sprang to the nearest trunk, only to fall back to the ground. The coordination needed for digging her claws into the bark was not quite back to normal. She paused briefly to recover from this surprise and glanced over toward the biologists. They had not moved. For the first time she was aware of the radio-collar and tugged at it with her forefeet. It stayed put. Again, she made a leap up the base of the tree and this time her claws held. Up the trunk she galloped, now feeling like her old self, and did not pause until well up in the crown. Here she stopped to peer back down. There were the biologists, watching her through binoculars. Soon they gathered up their equipment and walked off through the trees. Linea shrugged, tugged once more at her collar, then settled down in a comfortable crotch to regain her aplomb.

Chapter Seven

Telemetry's Brief Reign

Two days later, Linea poked her head tentatively out of her nest. It would be another half hour before sunup. Yet, even this early, she caught sight of the biologist leaning against a nearby ponderosa, his field glasses trained squarely on her. At his feet lay a small radio receiver with its ungainly antenna. Linea used several nests - how had he known she would be in this particular one, hours since her last activity? The all-but-forgotten collar around her neck was the key to this riddle, as it emitted its unrelenting signals. Imperceptible to any of her senses, even as they passed easily through her own flesh and bones, they were meaningless pulses of energy except when electronically translated into audible beeps. At any hour of the day or night, a person could rotate the attached antenna to the direction of loudest reception and eventually locate the signal's source... much as a bloodhound follows a trail of scent.

Eventually Linea descended and went about her usual routines, the biologist following at a great enough distance to minimize any influence his presence might have, but close enough to keep track of her through the glasses. In the following days, the procedure became fairly predictable. Linea was followed for a couple of hours at different times of day and became accustomed to the persistent but harmless attention. No longer were there any baited traps to divert her from natural pursuits. In a painfully slow process, a picture of Linea's 35-acre home range began to emerge and it soon became apparent that she spent much more time in some locations than in others. These "hot spots" posed tantalizing questions since, for the most part, they were not correlated with any features of the environment so far as the biologist could discern. No doubt Linea could have explained these and other nuances of her activity to him with the greatest of ease (and, probably, condescension)! But this would have required a type of communication not known to either.

One early afternoon, Linea rested on her stomach on a horizontal limb, all four feet and tail dangling down. She cast an apathetic eye toward the biologist who was watching her from one of his habitual posts.

Rarest mutant of any tassel-eared squirrel, this white-bellied Kaibab squirrel is seldom encountered. Neither the significance of belly color nor tail color is known.

Tom and Pat Leeson

Suddenly, movement in a neighboring pine caught her attention. It was another Kaibab squirrel, a total stranger. Not only was he a stranger to her, but his color pattern was unlike other Kaibab squirrels. The fur of his belly and inner legs was completely white, like the Abert's squirrels across the Grand Canyon on the South Rim. But the stranger's tail was identical to hers — not gray like that of an Abert's squirrel. In a few minutes the biologist also caught sight of this mutant creature and jumped to his feet in amazement. To him, infinitely more than to Linea, this was something new under the sun! The newcomer, apparently not noticing Linea, moved deliberately along, occasionally on the ground, then among the tree tops with the agitated biologist not far behind.

During the following week, Linea frequently encountered Whitebelly, but at this season of the year they paid little attention to each other. He established a home range that broadly overlapped hers and the biologist was torn between locating her by the ear of telemetry and locating him by his unique coloration.

Then a dramatic turn of events tipped the balance in favor of Linea, in the sense that her unwitting cooperation in the telemetry venture came to an abrupt halt. Although she had long ago become almost oblivious to the collar, she still worried it at times. She would give it a tug or two, then forget it and move on. On this particular day she chanced to try a new maneuver — she pushed it upward with both forepaws at once. In his effort not to cinch Linea's collar too tightly on the day she was trapped, the biologist had overcompensated slightly. When Linea now pushed it up, the collar barely slipped past her ears, then easily past her eyes and nose, dropping completely off. At the time, she was perched next to the trunk on a short, dead stub about fifteen feet above ground. By an incredible coincidence, the collar landed squarely on the stub, in perfect balance. Linea turned and descended the tree, leaving the collar to continue transmitting from its lonely perch. It pulsed all that afternoon and pulsed all night. It was still pulsing the next morning when the biologist appeared at the northern edge of Linea's home range, telemetry receiver in hand. He picked up the clear beeps as usual and finally zeroed in on the tree deserted almost a day ago by the squirrel. Looking up the trunk for a familiar white tail, he saw nothing. He then made a complete circle around the tree, hoping to see a nest accidentally missed previously. No luck.

Thoroughly confused, he made another circle, carefully monitoring the strength of the signals and confirming that they were indeed coming from the tree. In frustration, he surveyed the ground at the base, then looked up again. There it was — a suspiciously unnatural object balanced on a dead stub! A glance through the field glasses confirmed its identity.

Discovering the transmitter, jettisoned so precisely, was a bitter pill for the biologist. The temptation to view this as an act of premeditated spite (although he well knew it could not have been) was almost irresistible. Not having the heart to put Linea through a second ordeal of live-trapping and anesthetic, he resolved to continue observing her and Whitebelly, relying solely on their natural field marks.

Radio collar for a Kaibab squirrel. Shown life-size.

Chapter Eight

By Dusk and by Dark

Seldom seen because it is nocturnal and well-camouflaged, the common poorwill may be the only bird that hibernates in cold weather.

As the light of early evening began to fail, Linea settled into her nest for the night. It was one of her most predictable routines. Now, when squirrels and the host of other day creatures retire at dusk, others begin to bestir themselves. Hardly had the sun set behind Mount Trumbull and the first stars begun to appear when an ethereal vesper, "poor-WILL, poor-WILL, poor-WILL," broke the stillness. Not a forest sound, it was wafted up from the manzanita slopes of Shiva Saddle, a thousand feet below the rim at Tiyo Point. Its author, a large-eyed, mottled brownish bird a bit smaller than a robin, is named poorwill for its call. The bird crouched on the dry, rocky ground in an opening between shrubs as it performed. So beautiful is its camouflage that the poorwill usually goes unnoticed, even in broad daylight. Like the swift, it feeds on insects caught on the wing. But its style is as different from that of a swift as its schedule is. The one a consummate aerialist, the other almost earth-bound by comparison. As a moth passed by only a few feet above the slope, the poorwill interrupted its song to launch itself in pursuit. Its wide and rounded wings, gave the bird itself the appearance of a large moth. In fact, part of its scientific name, *Phalaenoptilus*, translates as "moth-feathered." After a short pursuit, the poorwill succeeded in engulfing the prey in its huge, bristle-rimmed mouth, then settled again on the ground. Once more its haunting song filled the still air.

A stone's throw from Linea's nest tree stood a half-dead aspen, the same tree from which she had gleaned shredded bark to line her nest weeks ago. Much of its bark was peeling away from the trunk in fraying plates. As the poorwill's notes floated up from the depths, a small bat emerged from beneath one of these plates of dead bark. Clinging upside down momentarily at the entrance to its roost, the bat cocked its head forward and scanned the surroundings. Its erect, blackish ears were several times longer than wide, contrasting obviously with the tan body. Tiny black eyes peered out from the surrounding fur of an elfin face. This, a long-eared myotis, is one of the commonest "mouse-eared" bats of the North Rim. Abruptly, the little bat released its toe grip, unfolded wings of bare, wrinkled skin, and was airborne. Off it flew, in the erratic flight peculiar to bats, for Harvey Pond. Here it coursed back and forth in long, straight swoops, just barely above the surface. On each traverse, it briefly dipped its lower jaw to scoop up a few drops of water, its wing tips dimpling the surface on each side. Thirst quenched, Myotis flew off to begin an evening of foraging. Through the cool darkness it danced and darted, mouth agape, now weaving expertly between pine branches, now swooping along only a few feet above the ground.

Myotis was now utterly dependent on sound. The loud, piercing cries he gave in flight, too high pitched for humans to hear, bounced back as echoes from all the fabric of his surroundings. And the echoes, each subtly altered according to whether it returned from trunks, twigs, rocks, forest floor or airborne objects, were relayed from frequency of his cries to obtain still finer resolution. Now Myotis was able to tell how far away the insect was, its size, speed and direction of flight, even the texture of its

Bats usually fly with their mouths wide open. This is not for panting or a sign of aggressiveness, but is for uttering their ultra-sonic cries. Echoes from passing insects guide the bats in capturing their prey. This species is the long-eared myotis.

Myotis' ears to his brain. This compact computer, smaller than a pea, instantly processed the stream of information in astonishing detail. Myotis could "see" all nearby objects as a sound-picture. Prime targets were flying insects. He swung left and began receiving echoes from a mosquito. Immediately he stepped up the body surface! As he closed in, following his prey's every twist and turn, Myotis prepared for the capture. In the split second before collision, he flipped down the membrane stretched between hind legs and tail to make a small parachute. In an amazing feat of coordination, he scooped up the mosquito and reached down to grab it in his teeth.

Without a pause in flight, Myotis chewed and swallowed en route. His next destination was a small clearing in the forest where he knew flying insects teemed at this time of night. Here his job was easier because fewer twigs and branches meant fewer obstructions to avoid in flight. But several bats had already beaten him to this feeding arena; others soon joined in, each intent on similar business. Before long Myotis was one of a dozen, swirling around the clearing, a flittering carousel of mouse-sized predators. Again and again, using his finely honed sonar, Myotis detected, pursued, and snatched up flying morsels, consuming them on the wing. However, his numerous competitors greatly complicated his efforts: all gave their own pulsing cries and echoes which somehow had to be distinguished from his own.

When one clearing became depleted, Myotis tried another. After an hour, he had caught more than two hundred insects, mostly mosquitoes, and it was time for a siesta.

Back he circled to the aspen roost, swooped up to his flake of bark, stalling precisely as he folded his wings to fit into the overhung crevice. Abruptly the high-pitched sonar snuffed out; in his narrow crypt Myotis hung by his feet and closed his eyes. Outside, the light of the rising moon shone on his retreat as a great horned owl's "Uh-who-who-whooo" sounded from the rim. Again the poorwill's call drifted up from Shiva Saddle. Moonlight bathed Shiva Temple, exaggerating its isolation from the rest of the Canyon beyond. There it seemed to float, a monolithic dreadnought, serenely aloof from passing concerns of poorwills, bats and squirrels.

Chapter Nine

Storm, Fire and Sunset

Amanitas with a nibble mark.

By mid-August the monsoon season was in full swing. A typical morning started with a perfectly clear sky. Now that the breeding season was over and bird songs had dwindled to almost nothing, the forest seemed strangely silent. An hour or two before noon, puffy white, innocent looking cumulus clouds began to appear; by early afternoon they were piling up into ominous thunderheads. Soon a few heavy, scattered rain drops began to fall, and lightning and thunder ushered in a dramatic but short-lived storm. Heavy rain, often alternating with hail, drove life under cover. As the afternoon wore on, the storm lost its momentum, the rain abated and the cloud cover broke up and scudded away. It was a favorite time of day for Linea to forage in the water-soaked duff for the various mushrooms now appearing in profusion. The abundant boletes, inky caps and amanitas were easy to find. She took a few bites from an amanita, so poisonous to humans, then stored the remainder in a low crotch of a nearby pine. But her favorites were the truffles. Just as abundant as the others but completely invisible a few inches below the surface, these white, globular fungi had to be located by scent, then dug out one by one. Each truffle, as it developed in its dark, subterranean niche, had established an intimate connection with the nearest pine before being rudely excavated by Linea. Microscopic rootlike threads growing down from the fungus envelope the delicate rootlets of the tree as a mantle and extend from the mantle into the soil. This mantle, penetrating root tissue, forms a network that helps transfer nutrients and water from the soil to the root tissues that lead to other parts of the tree. Conversely, the root tissues act as storage sites for carbohydrates and amino acids which are passed on to nourish the fungus. It is a mutually beneficial partnership. Linea's role as a consumer of the pine's male and female cones, plus its sugary inner bark and, finally, its hidden fungal partners was not completely destructive. For, along with the nutritious body of the fungus, she also consumed the completely indigestible spores. These were later passed out, often hundreds of yards away. Thus, by serving as a disperser of spores,

she unwittingly performed a crucial service for the pine community upon which her entire life style depended. Above ground fungi like amanitas produce spores that are carried far and wide by the wind. But the inch or two of soil that gives one truffle generation some protection, deprives it of pioneering new sites for the next generation. Unless, of course, its spores can hitch a ride inside a passing squirrel or deer.

Tree crotches are commonly used sites for temporary storage of mushrooms.

Squirrels are quite fond of these hypogeous (underground) mushrooms which they locate by scent and excavate. They are important for mushroom dispersal because the spores pass through their digestive systems and are deposited far and wide.

An enormous pile of clouds hung over the Walhalla Plateau one early August afternoon. A jagged flash of lightning darted to the ground. After almost a half minute, there was a muted rumble of thunder. Scouting about for buried truffles, Linea paid no heed. The ever-growing gray mountain of thunderheads advanced inexorably westward across Bright Angel Canyon and the time lag between seeing a flash and hearing the resulting clap grew shorter and shorter. As the rain began, Linea finally looked up somewhat anxiously from where she sat on her haunches nibbling on a truffle. Across the glade she noted Whitebelly climbing one of his nest trees. She decided to wait out the downpour on the protected snag of a nearby ponderosa. Up she scrambled, hunched over next to the trunk where an overhanging branch sheltered her. Her tail, arched over her back, also helped. For a few minutes she stuck it out, the rain dripping off her tail, back, and sides. But then the rain fell much harder, a bright stab of lightning lit up the forest, and a deafening clap of thunder followed without noticeable delay. Alarmed, Linea scrambled down to the ground, galloped to her own nest tree as fast as she could and dashed up to the shelter of the nest. Hardly had she entered before a cataclysmic bolt of lightning struck so close that even the inside of her nest was lit up in a violet glow for a split second. At the same instant came a deafening blast. This was caused not only by the thunder clap but also by the explosion of the pine standing next to her nest tree. The lightning bolt, much more powerful than most, had made a direct strike, vaporizing the sap in the trunk so violently that the tree exploded, blowing out large fragments in all direction, like javelins. Linea huddled in her swaying chamber; in a few seconds she heard a crackling and smelled smoke as what remained of the tree's crown caught fire. It began to blaze like a torch as the resinous needles ignited. Luckily, the flames did not leap across to ignite the crown of Linea's tree. But several burning branches fell and started a fire on the ground. This slowly grew out in a circular pattern, consuming the foot-high understory of fir but sparing most of the less vulnerable pine seedlings. Meanwhile, the heat from the burning crown

Periodic fires are necessary for optimum growth of ponderosa pines. Crown fires, right, often resulting from too much fuel accumulation on the ground, can be quite destructive. On the other hand, ground fires, above, keep young firs and other competitors in check, sparing the pines.

was making Linea's nest intolerable. She looked out her doorway and saw the chaos of flames. In panic she withdrew. But then, driven to desperation by the heat, she bolted out and raced down to the ground. It must surely have been more terrifying than being caught in a live trap!

Fortunately, the ground fire had not reached her tree, so a wide swath of unburned duff provided an avenue for escape. She ran in the direction of another nest-tree and was nearly trampled by the hooves of a fleeing buck. There was Whitebelly bounding along ahead of her. Soon both were beyond the fire. The storm began to clear, and the rain tapered off. Safe in another nest where one of her daughters had joined her, Linea finally dozed as late afternoon gave way to early evening. Both squirrels roused themselves for a final mushroom snack. Although the fire had died down, a pall of smoke hung in the air. By the time they had made their descent, the sun was about to set and Linea had almost forgotten her ordeal by fire. Yet, paradoxically, the seemingly destructive phenomenon of fire was, in the long run, essential in the maintenance of this forest by discouraging fir seedlings and other undergrowth.

A few aspens were beginning to show golden leaves, quiet testimony that another summer was phasing into fall. The last rays of the setting sun glanced along the ground, backlighting sparse heads of grass. As Linea cantered leisurely across the opening between two long shadows, her tail caught the light and for a few lingering seconds, had a bright ember glow of its own. It was one of those rare and fleeting images, so stunning that it transcended the immediate. To the observer, it was no mere snapshot from the life of one individual. Rather, it was symbolic - a timeless reference to the lives of *all* Kaibab squirrels.

Chapter Ten

Postlude to Summer

It was as if the fire had been the cue for some shock-wave flung across the Kaibab Plateau. A week later almost all the aspens had turned a brilliant yellow and every grove seemed to bask in its own sunlight; crimson splashes on the slopes above the Coconino sandstone advertised each stand of Rocky Mountain maple and russet patches highlighted the groves of Gambel's oak.

Linea felt more energetic now that there was a nip to the air and she and Whitebelly could be seen harvesting pine cones with new enthusiasm. Both had just regrown most of their ear tassels, a sure sign of autumn.

Ear tassels announce the arrival of autumn.

Another sign - the mule deer were less retiring nowadays. Fawns had lost their spots and were consorting with the adults. The bucks, necks beginning to swell, were

boldest of all. Two of them, one in its prime, the other with a couple of years still to go, moved sedately across a clearing, then beneath the tree where Linea perched. The smaller buck's antlers were already clean and polished but some tatters of velvet still dangled from the great rack of the patriarch. Suddenly, both raised their heads and gazed intently over their shoulders. A whiff of some scent, probably that of a mountain lion, alarmed them. Giving a couple of sharp snorts, both bolted up the oak-locust slope in that peculiar "pronking" gait in which all four hooves strike the ground in unison, then lift off in unison. Both squirrels froze in place for a few minutes, their usual reaction to any hint of danger, then resumed feeding.

By early afternoon Linea and Whitebelly had eaten their fill and were content to pause and sleepily survey the scene below. Across the far side of the clearing a coyote appeared amongst the bracken ferns and trotted toward an aspen stand. Even the most casual observer of coyotes would have to agree that they have a trot that is in a class by itself. They move so smoothly, so effortlessly that they give the impression of not working at all. Rather, they appear to be propelled by some external force - to miraculously float across the landscape! What amazing grace.

As the coyote entered the grove of aspen, a gust of wind came up, releasing a shower of golden leaves. He glanced up at this spectacle briefly, then at the large ponderosa pine beyond. On a high branch he picked out two motionless, dark gray forms, both with white tails. One of them had a white belly. Without breaking his stride, he lowered his gaze, continued on his way and disappeared.

Now the breeze subsided and a few last aspen leaves tumbled down from above, as song-notes of warbling vireos had done four months ago.

Joseph Hall is professor of biology emeritus at San Francisco State University and a fellow and research associate at the California Academy of Sciences.

Born in southern Ohio, he spent much of his boyhood wandering through local woods, fields and wetlands observing wildlife. Armed with a Brownie box camera, he also began what would be a lifelong hobby of nature photography.

In his teens, a family trip to the West made an indelible impression on him and he has lived in California for most of his professional life. The Grand Canyon was particularly enthralling and he worked seasonally for the National Park Service there, first as a Ranger-Naturalist and later as a Collaborator.

In addition to his work on the Kaibab squirrel, he has made intensive studies of beavers in California and river otters in Wyoming.

He and his wife Betty now live in Grand Junction, Colorado.